Table of Contents

Catastrophism 3

GEOLOGIC CATASTROPHES 3

Let's begin with Mega Floods. 4

Impact of Asteroids on the East coast of North America 7

Sea Level Rise 11

The Earth's natural climate cycle 21

Solar influences 22

Trends in stratospheric temperature since systematic measurements began, from eight different datasets. {Source: The Earth's natural climate cycle 24

Solar influences...... 25

Worldwide ice sheets 29

ICE SHEETS 30

Defining an epoch 32

Life during the Ice Age 33

S0, how do we react? 35

Wisdom Ignored: 40

Other Books by Vernon Finney 49

Catastrophism

Theory that the Earth has frequently been altered by sudden, short-lived violent events!

GEOLOGIC CATASTROPHES

Natural catastrophes include floods, rapid uplift of mountains, earth quakes tornadoes, Tsunamis, Cyclones, changes in Ocean Currents, and Atmospheric Circulation!

This Book Will Discuss three of these issues, With Special Emphasis on Sea Level Rises!

Let's begin with Mega Floods.

Source NOVA DVD!

Flying westward over the Dakotas one can see 300 ft high Ripple Marks.

A Geologist was scorned by his

colleagues for years for his explanation of these Ripple Marks! Finally, his explanation of the failure of a huge glacial dam was accepted! His further studies confirmed that such dam failures had occurred frequently!

Impact of Asteroids on the East coast of North America?

This photo shows flood marks at 400 feet above Sea Level. Geologic findings have confirmed such sea level rises in the past! Caused by typhoons,

Asteroids, etc. and predicted in the near future!

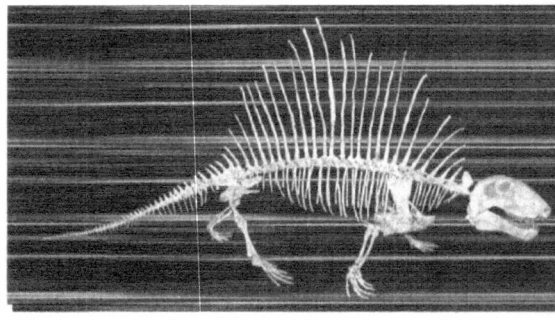

Dimeterdons were among the many species killed off by the Permian-

Triassic extinction event! Of all the mass extinctions that have devastated our world; the Permian-Triassic was the worst! No coal deposits were formed in the early Triassic!

The date is Dec. 1. 2029, 2019. Where are

we in Environmental Actions?

Let's start with Global Warming! What should be a given, continues to be doubted by many: The melting of glacial ice in the Artic and Antarctica is well documented. The less ice there is, the more open water there is to absorb heat and the less ice there is to reflect sun light! The heat rising from the warm waters rises into the atmosphere! Circulatory patterns are changed by this rising heat and causes changes in seasonal conditions worldwide!

So, to question Global Warming due to cold fronts in lieu of warmer conditions simply shows ignorance. I direct doubters to books like "Global Warming For Dummies by Elizabeth May".

Of even more sincere concern is the lack of discussion of Catastrophic Events that may occur and have been documented in Ice Cores! So, let's address Sea Level Rise.

Sea Level Rise

The question should not be whether sea level is rising but

how fast may sea level rise occur.

Sea level rises and falls have occurred rapidly in the last 15,000 years and are likely to occur again. Weather patterns can permanently shift in as little as a year, according to the records preserved in an ice core from Greenland. Source: Scientific American, By David Biello| June 23, 2008 |

Like many other cities, Norfolk VA was built on filled-in marsh. Now, that fill is settling and compacting. In addition, the city is in an area where significant natural sinking of land is occurring. The result is that Norfolk has experienced the highest relative increase in sea level on the East Coast — 14.5 inches since 1930, according to readings by the Sewell's Point naval station.

Approximately 5 inches of the 14.5 inches has been attributed to sea level rise

The continental glaciers of Antarctica

have been shown to rest on bed rock separated by a thin layer of super cooled water that behaves like a lubricant. Shelving of glacial lobes is evidence of the glacial mass moving into the surrounding Ocean. If the entire glacial mass on Antarctica should suddenly slide into the surrounding Ocean, sea level could rise 30 plus feet. There is evidence in glacial cores that this has occurred in the geologic past. **The only salvation from such a catastrophic**

event is if ablation of the overlying glacial mass is occurring fast enough to alleviate the mass required for such a catastrophic occurrence!

The planet is warming, from North Pole to South Pole. Since 1906, <u>the global average surface temperature has increased by more than 1.6 degrees</u> Fahrenheit (0.9 degrees Celsius)—<u>even more</u> in sensitive polar regions. And the impacts of rising temperatures aren't waiting for some far-flung future–the effects of global warming are appearing right now. The heat is melting glaciers and sea ice, shifting precipitation patterns, and setting animals on the move.

Many people think of global warming and climate change as

synonyms, but scientists prefer to use "climate change" when describing the complex shifts now affecting our planet's weather and climate systems. Climate change encompasses not only rising average temperatures but also extreme weather events, shifting wildlife populations and habitats, <u>rising seas</u>, and a range of other impacts. All of these changes are emerging as humans continue to add heat-trapping <u>greenhouse gases</u> to the atmosphere!

- Ice is melting worldwide, especially at the Earth's poles. This includes mountain glaciers, ice sheets covering West Antarctica and Greenland, and Arctic sea ice. In Montana's Glacier National Park the <u>number of glaciers has declined</u> to fewer than 30 from more than 150 in 1910.
- Much of this melting ice

contributes to <u>sea-level rise</u>. Global sea levels are <u>rising 0.13 inches (3.2 millimeters) a year</u>, and the rise is occurring at a faster rate in recent years.
- Rising temperatures are affecting wildlife and their habitats. Vanishing ice has challenged species such as the <u>Adélie penguin in Antarctica</u>, where some populations on the western peninsula have collapsed by 90 percent or more.
- As temperatures change, <u>many species are on the</u>

move. Some butterflies, foxes, and alpine plants have migrated farther north or to higher, cooler areas.
- Precipitation (rain and snowfall) has increased across the globe, on average. Yet some regions are experiencing more severe drought, increasing the risk of wildfires, lost crops, and drinking water shortages.
- Some species— including mosquitoes, ticks, jellyfish, and crop pests— are

thriving. [Booming populations of bark beetles](#) that feed on spruce and pine trees, for example, have devastated [millions of forested acres](#) in the U.S.

Climate Cycles:

The Earth's natural climate cycle

Over the last 800,000 years, there have been natural cycles in the Earth's climate. There have been ice ages and warmer interglacial periods. After the last ice age 20,000 years ago, average global temperature rose by about 3°C to 8°C, over a period of about 10,000 years.

We can link the rises in temperature over the last 200 years to rises in atmospheric CO_2 levels. Rises in temperature are now well above the natural cycle of the last 800,000 years.

Solar influences

The sun is the primary source of Earth's heat, so

relatively small changes in solar output can affect our climate.

Satellite observations since the late 1970s have shown a slight decrease in the sun's total energy output. However, instead of cooling, the Earth has warmed over this period.

Also, warming from the sun would heat all of the atmosphere, including the lowest few kilometres (the troposphere) and the layer above (the stratosphere).

Observations show that the stratosphere is in fact cooling while the troposphere warms. This is consistent with greenhouse gas heating and not solar heating.

Trends in stratospheric temperature since systematic measurements began, from eight different datasets. {Source:

The Earth's natural climate cycle

Over the last 800,000 years, there have been natural cycles in the

Earth's climate. There have been ice ages and warmer interglacial periods. After the last ice age 20,000 years ago, average global temperature rose by about 3°C to 8°C, over a period of about 10,000 years.

We can link the rises in temperature over the last 200 years to rises in atmospheric CO_2 levels. Rises in temperature are now well above the natural cycle of the last 800,000 years.

Solar influences

The sun is the primary source of Earth's heat,

so relatively small changes in solar output can affect our climate.

Satellite observations since the late 1970s have shown a slight decrease in the sun's total energy output. However, instead of cooling, the Earth has warmed over this period.

Also, warming from the sun would heat all of the atmosphere, including the lowest few kilometres (the troposphere) and the layer above (the stratosphere). Observations show that the stratosphere is in fact cooling while the troposphere warms. This is consistent with greenhouse gas heating and not solar heating.

Trends in stratospheric temperature since systematic measurements began, from eight different datasets.

Source:

usCommittee on Climate Change
151 Buckingham Palace Rd,
London,
SW1W 9SZ

The Pleistocene Epoch is typically defined as the time period that began about 2.6 million years ago and lasted until about 11,700 years ago. The most recent Ice Age occurred then, as glaciers covered huge parts of the planet Earth.

There have been at least five documented major ice ages during the 4.6 billion years since the Earth was formed — and most likely many more before humans came on the scene about 2.3 million years ago.
The Pleistocene Epoch is the first in which *Homo sapiens* evolved, and by the end of the epoch humans could be found in nearly every part of the planet. The Pleistocene Epoch was the first epoch in the Quaternary Period and the sixth in the Cenozoic Era. It was followed

by the current stage, called the Holocene Epoch.

Worldwide ice sheets

At the time of the Pleistocene, the continents had moved to their current positions. At one point during the Ice Age, sheets of ice covered all of Antarctica, large parts of Europe, North America, and South America, and small areas in Asia. In North America they stretched over Greenland and Canada and parts of the northern United States. The remains of glaciers of the Ice Age can still be seen in parts of the world, including Greenland and Antarctica

ICE SHEETS

But the glaciers did not just sit there. There was a lot of movement over time, and there were about 20 cycles when the glaciers would advance and retreat as they thawed and refroze. Scientists

identified the Pleistocene Epoch's four key stages, or ages — Gelasian, Calabrian, Ionian and Tarantian.

The name *Pleistocene* is the combination of two Greek words: *pleistos* (meaning "most") and *kainos* (meaning "new" or "recent"). It was first used in 1839 by Sir Charles Lyell, a British geologist and lawyer. As a result of Lyell's work, the glacial theory gained acceptance between 1839 and 1846, and scientists came to recognize the existence of ice ages. During this period, British geologist Edward Forbes aligned the period with other known ice ages. In 2009, the International Union of Geological Sciences established the start of the Pleistocene Epoch at 2.588 million years before the present.

Defining an epoch

Overall, the climate was much colder and drier than it is today. Since most of the water on Earth's surface was ice, there was little precipitation and rainfall was about half of what it is today. During peak periods with most of the water frozen, global average temperatures were 5 to 10 degrees C (9 to 18 degrees F) below today's temperature norms.

There were winters and summers during that period. The variation in temperatures produced glacial advances, because the cooler summers didn't completely melt the snow.

Life during the Ice Age

While *Homo sapiens* evolved, many vertebrates, especially large mammals, succumbed to the harsh climate conditions of this period.

One of the richest sources of information about life in the Pleistocene Epoch can be found in the La Brea Tar Pits in Los Angeles, where remains of everything from insects to plant life to animals were preserved, including a partial skeleton of a female human and a nearly complete woolly mammoth.

In addition to the woolly mammoth, mammals such as saber-toothed cats (Smilodon), giant ground sloths (Megatherium) and mastodons roamed the Earth during this period. Other mammals that thrived during this period include moonrats, tenrecs (hedgehog-like creatures) and macrauchenia (similar to a llamas and camels). {Source: Kim Ann Zimmerman August 29,2017 History}

S0, how do we react? We need to plan ahead for the changes we can predict. Agriculture needs to change to most favorable areas for crop growth. Sea Walls need to be built to protect coastal areas!

All this can be achieved if we place faith in our Scientists (Science being a gift of God)

History is full of examples of failure to listen to Science:

The advent of agriculture was the first significant impact of man on mother earth. Clearing of natural vegetation, draining of marshes significantly began to alter the world in which man lives. Evidence suggests that concentrations of CO_2 started rising about 8,000 years ago, even though natural trends indicate they should have been dropping. Some 3,000 years later the same thing happened to methane, another heat-trapping gas. The consequences of these surprising rises have been profound. Without them, current temperatures in northern parts of North America and Europe would be cooler by three to four degrees Celsius--enough to make agriculture difficult. In addition, an incipient ice age--marked by the appearance of small ice caps--would probably have begun several thousand years ago in parts of northeastern Canada. Instead the earth's climate has remained relatively warm and stable in recent millennia.

Source:**How Did Humans First Alter Global Climate?** ByWilliam F. Ruddiman, Scientific

American | February 21, 2005 |

Ice Malting And Sea level rise

- Sea ice influences climate because it reflects sunlight and because it influences ocean circulation.
- Less sea ice leads to acceleration of global warming
- There is evidence of ice melt, sea level rise to +5-9 m, and extreme storms in the prior interglacial period that was less than 1°C warmer than today.
- Arctic sea-ice cover is shrinking by 8.9% per decade in summer and 2.5% per decade in winter. It is also becoming thinner and there is less multi-year ice.
- Melting sea ice, in combination with melting glaciers and ice sheets, may cause major changes to global patterns of ocean circulation.
- As with snow, less sea ice increases absorption of heat from the sun, resulting in increased warming

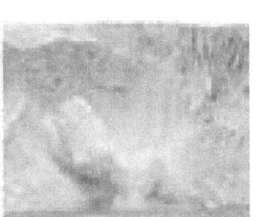

Wisdom Ignored:

In 1932 Hugh Hammond Bennett toured the High Plains just as the soil began to blow! His diagnosis was it was not natures fault! The cause was man. Indigenous people had farmed for centuries and not lost the soil. In barely a generation Americans had stripped the land of its life-giving layers.

Americans had changed the face of the land more than the combined activities of volcanoes, earthquakes, tidal waves, tornadoes, and all the excavations of mankind since the beginning of history.

In 1933 Aldo Leopold invoked "The Conservation Ethic!

For the land to be restored in

the manner obvious to XIT

cowhands and the

Comanches:

Grass for animals that eat grass! The struggle went on until Bennett put Roosevelt's agency the Civil Conservation Service (CCC) to work.

CCC

The Civilian Conservation Corps (CCC) was a public work relief program that operated from 1933 to 1942 in the United States for unemployed, unmarried men. Originally for young men ages 18–25, it was eventually expanded to ages 17–28.[1]Robert Fechner was the first director of the agency, succeeded by Mcanteen following Fechner's death. The CCC was a major part of President Franklin D. Roosevelt's New Deal that provided unskilled manual labor jobs related to the conservation and development of natural resources in rural lands owned by federal, state, and local governments. The CCC was designed to provide jobs for young men and to relieve families who had difficulty finding jobs during the Great Depression in the United States. Maximum enrollment at any one time was 300,000. Through the course of its nine years in operation, 3 million young men participated in the CCC, which provided them with shelter, clothing, and food, together with a wage of $30 (about $570 in 2017[2]) per month ($25 of which had to be sent home to their families).[3]

The American public made the CCC the most popular of all the New Deal programs.[4]Sources written at the time claimed[5] an individual's enrollment in the CCC led to improved physical condition, heightened morale, and increased employability. The CCC also led to a greater public awareness and appreciation of the outdoors and the nation's natural resources, and the continued need for a carefully planned, comprehensive national program for the protection and development of natural resources.[6]

Enrollees of the CCC planted nearly 3 billion trees to help reforest America; constructed trails, lodges, and related facilities in more than 800 parks nationwide; and upgraded most state parks, updated forest fire fighting methods, and built a network of service buildings and public roadways in remote areas.

Despite its popular support, the CCC was not a permanent agency. It depended on emergency and temporary Congressional legislation and funding to operate. By 1942, with World War I and the draft in operation, the need for work relief declined, and Congress voted to close the program.[10]

SCS

On April 27, 1935 Congress passed Public Law 74-46, in which it recognized that "the wastage of soil and moisture resources on farm, grazing, and forest lands . . . is a menace to the national welfare," and it directed the Secretary of Agriculture to establish the Soil Conservation Service (SCS) as a permanent agency in the

USDA. In 1994, Congress changed SCS's name to the Natural Resources Conservation Service (NRCS) to better reflect the broadened scope of the agency's concerns.

The creation of the Soil Conservation Service represented the culmination of the efforts of Hugh Hammond Bennett, "father of Soil Conservation" and the first Chief of SCS, to awaken public concern for the problem of soil erosion. Bennett became aware of the threat posed by the erosion of soils early in his career as a surveyor for the USDA Bureau of Soils. He observed how soil erosion by water and wind reduced the ability of the land to sustain agricultural productivity and to support rural communities who depended on it for their

T

livelihoods. He launched a public crusade of writing and speaking about the soil erosion crisis. His highly influential 1928 publication "Soil Erosion: A National Menace" influenced Congress to create the first federal soil erosion experiment We have much to learn from the Indigenous Peoples of the world, who know how to live with nature and in carrying out God's Gift of

T

Making Us Caretakers of

HIS World!

Other Books by Vernon Finney

***The World According to Vern a series of five;
Three Toes***

T

Lives On; The First Environmental Engineers, Ham, Pork and Chop; Oso Gigante, Big Red Mad As A Hatter, Follow

The Whales, Berengia to Tierra del Fuego, Who Am I Lord, Chulito and The Leprecaun

T

Greed VS GOD
Climate

www.ingramcontent.com/pod-product-compliance
Lightning Source LLC
Chambersburg PA
CBHW030531220526
45463CB00007B/2785